This book is the result of a con .... with a friend following my return from having spoken at the opening session of the Paris Climate Change Summit that took place in December 2015.

He suggested that most people really don't understand what climate change is all about. "Why don't you produce a simple plain English guide to the subject?" he asked. Having thought about it, I began to realize that this might perhaps be rather a good idea, considering how it is a subject that should be of profound disquiet to anyone concerned about the long-term future of our world.

After discovering that Ladybird Books were indeed interested in producing such a guide in their "Ladybird Expert Series", I turned to Tony Juniper (whose knowledge on the subject is prodigious) and to Emily Shuckburgh (a leading climate scientist) for assistance in compiling the book before you now. To ensure the accuracy of our collective effort the book has been subject to peer-review by eminent academics coordinated under the aegis of the Royal Meteorological Society.

I hope this modest attempt to alert a global public to the "wolf at the door" will make some small contribution towards encouraging requisite action; action that must be urgently scaled up, and scaled up <u>now</u>.

Series 117

This is a Ladybird Expert book, one of a series of titles for an adult readership. Written by some of the leading lights and outstanding communicators in their fields and published by one of the most trusted and well-loved names in books, the Ladybird Expert series provides clear, accessible and authoritative introductions, informed by expert opinion, to key subjects drawn from science, history and culture.

The Publisher would like to thank the following for the illustrative references for this book:
Cover and page 11 from photo © Wittwoophoto/Caters News; page 9 from photo © Wang Chengyun, Xinhua/Landov/Barcroft Media; page 13 from photo © Gary Braasch; page 19 from original Ladybird illustration by John Kenney; page 21 British Antarctic Survey © BBC; page 25 from original Ladybird illustration by Frank Humphris; page 29 from photo © Robert Mulvaney; page 33 Marshal Islands from photo © Reinhard Dirscherl/Getty Images; page 43 from original Ladybird illustration by Charles Tunnicliffe; page 46 from photo © Arnaud Bouissou-Medde/SG COP21/Flickr and Getty Images.

Every effort has been made to ensure images are correctly attributed, however if any omission or error has been made please notify the Publisher for correction in future editions.

MICHAEL JOSEPH

UK | USA | Canada | Ireland | Australia
India | New Zealand | South Africa

Michael Joseph is part of the Penguin Random House group of companies
whose addresses can be found at global.penguinrandomhouse.com

Penguin
Random House
UK

First published 2017
001

Text copyright © HRH The Prince of Wales, Tony Juniper,
Emily Shuckburgh, 2017

All images copyright © Ladybird Books Ltd, 2017

The moral right of the authors has been asserted

Printed in Italy by L.E.G.O. S.p.A.

A CIP catalogue record for this book is available from the British Library
ISBN: 978–0–718–18585–5

www.greenpenguin.co.uk

MIX
Paper from
responsible sources
FSC® C018179

Penguin Random House is committed to a
sustainable future for our business, our readers
and our planet. This book is made from Forest
Stewardship Council® certified paper.

Climate Change

**HRH The Prince of Wales
Tony Juniper
Emily Shuckburgh**

with illustrations by
Ruth Palmer

Ladybird Books Ltd, London

The Earth's climate

The climate has a profound influence on all our daily lives and has shaped the history of life on Earth. Climatic conditions are determined by the atmosphere, oceans, land, ice and the life on our planet acting in concert under the power of the Sun.

Earth's atmosphere forms a layer as thin in relative terms as the skin of an apple. It is mostly made up of nitrogen and oxygen, but also contains smaller amounts of other gases. These include those commonly referred to as greenhouse gases which trap the Sun's heat and keep the Earth warm enough for life to flourish.

Satellites orbiting the Earth and monitoring stations on the ground show that concentrations of greenhouse gases in the atmosphere are increasing, in particular carbon dioxide. This is leading to rising temperatures and disruption to the climate.

We are already seeing dramatic impacts: altered weather patterns, reduced snow and ice and a rise in sea levels.

These impacts threaten food and water supply, people's health, security and economic activity, as well as wildlife and the natural world. If we act now to tackle climate change, we will support progress towards a more prosperous, secure and sustainable future. But if we don't act immediately, this could all be at risk.

A warming world

Records from thousands of weather stations across the world, and ocean data from ships and buoys, show the temperature measured at the Earth's surface has increased substantially over the past century, and especially over the last fifty years. Certain regions, in particular the Arctic, have seen much more warming than others.

Some years have always been warmer or cooler than others. This is because natural factors can cause year-to-year and decade-to-decade temperature variations. These natural factors include changes to the strength of the Sun, the impact of volcanic eruptions and climate cycles such as the El Niño phenomenon.

Nevertheless, the three decades from 1980 to 2010 all showed record warmth compared with previous decades. Since the turn of the millennium, the world's temperature has typically been more than 0.75°C warmer than it was 150 years ago in Victorian times and in 2015 it reached 1°C warmer.

Other observations from around the world, including warming of the oceans from the surface to depths and reductions in ice and snow cover, provide further evidence that planetary-scale warming is taking place.

Global surface temperature

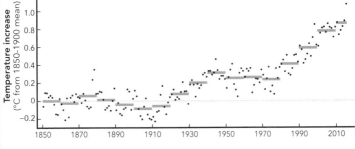

Melting ice and rising sea levels

The warming has dramatically reduced Arctic sea ice. In 2016 and several other recent years, at the end of the summer melt season, the sea ice covered an area less than two-thirds of that at the end of the twentieth century. Just to put that in perspective, that's a drop equivalent to the area of the United Kingdom, Ireland, France, Spain, Germany and Italy put together. The melting of this floating ice does not raise sea levels, but change on this scale can alter weather across Europe, Asia and North America.

In addition, as the world has warmed, the water in the oceans has expanded, many mountain glaciers have retreated and ice sheets in Greenland and Antarctica have shrunk. These changes *have* raised sea levels, as long-term measurements of tide gauges and recent satellite data show.

The effects of sea-level rise are felt most acutely when storm surges occur. Worryingly, many of the world's existing and developing megacities, including Shanghai, Jakarta and Mumbai, are located in vulnerable coastal regions. The flooding which hit New York City in 2012 during Hurricane Sandy showed the damage that storm surges can cause to critical infrastructure.

For now, the Thames Barrier protects London from flooding, but if significant sea-level rise occurs, expensive reinforcement will eventually be required. For many other cities, such defences would be either impossible or simply too costly to build.

Heatwaves, droughts, floods and storms

Extreme weather events such as heatwaves, droughts, floods and storms can cause major damage and disruption, with large financial costs and sometimes loss of life.

In 2016, India was reported to have recorded its then highest ever temperature (51°C) during a deadly heatwave. A severe summer heatwave in Europe in 2003 led to tens of thousands of premature deaths, especially among elderly people. In Australia, recent intense bushfires have destroyed thousands of properties and led to scores of deaths. Parts of the UK have repeatedly seen devastating flooding in the years since the start of the twenty-first century and in Pakistan floods affected 20 million people in 2010.

Around the world extreme weather conditions are leading to temperature and rainfall records being broken, with ever more serious consequences, as what were once extreme conditions are starting to become normal.

Freak weather has always occurred, but studies indicate climate change has increased the risk of certain extreme conditions, for example a major European heatwave. Analysis indicates that the kind of heavy downpours responsible for some of the terrible recent flooding in the UK have become more likely because of climate change. In part this is because a warmer atmosphere holds more water, giving rise in places to more intense rains and increased flood risk.

Threats to people and where they live

Threats to food and water supply, human health and national security, and the risk of humanitarian crises, are all potentially increased by climate change.

Crop yields depend on a range of factors including temperature, rainfall and sunshine. In the extreme, droughts and floods can push up prices to the point where the poorest go hungry. Some climate changes may favour certain crops, but overall it is expected that continued climate disruption will lead to less secure food supplies.

There is concern that the supply of water to some populations may become depleted as climate change advances. For example, across South Asia and China, large rivers provide fresh water for drinking and irrigation and the water flow depends on regional rain and snowfall, with meltwater from Himalayan glaciers providing additional supply. Each of these sources may be affected by climate change.

Human health can be affected by extreme weather as well as by factors associated with climate change such as air pollution and the spread of disease.

When climate change threatens basic human needs and welfare this can exacerbate existing tensions and may increase the risks of regional conflict and migration. Moreover, heatwaves, droughts and floods can complicate the relief effort for refugees displaced because of conflict.

Disappearing wildlife

Climate disruption is causing shifts in the conditions that sustain many wildlife species. This includes changes in temperature, the length and timing of the seasons, and the availability of water.

Under natural conditions, many species would be able to cope with gradual changes and adapt accordingly. Rapid climate changes along with the loss of natural habitats, mainly to agriculture and urbanization, create unprecedented challenges for wildlife.

Further climate change means some vulnerable populations of animals and plants will inevitably die out. For wildlife species that are already rare, declining or very specialized, climatic changes could lead to their extinction.

The polar bear has come to symbolize the threats posed to wildlife by climate change, but this magnificent animal is hardly alone. Across the world many other creatures are at risk, for example salmon, leatherback turtles, emperor penguins and beluga whales. A number of Australia's animals are under severe threat from habitat loss due to climate change, including the yellow-footed rock-wallaby, golden-shouldered parrot and Lumholtz's tree-kangaroo.

Many of our medicines and all the plants and animals we rely upon for food are derived from wild species. As the rate of extinction continues to gather pace, future generations may inherit a world that lacks the incredible wildlife diversity that we enjoy today and be poorer for it, not only economically, but also spiritually and culturally.

Impacts on businesses and communities

Some industries and communities are particularly vulnerable to the effects of climate change and are already paying a high price for weather disasters.

Insurance companies are entering an era in which it will be harder for them to provide protection against costly incidents like flooding of people's homes.

Food producers and water companies face increasing business risks arising from the effects of droughts, floods or other extreme weather. A recent extended drought in Texas, which peaked in 2011, cost farmers and ranchers billions of dollars in lost income.

Hydroelectric dams are being affected by severe droughts that reduce river flow, resulting in less water for power generation. For communities in Brazil this has been a major source of disruption.

All of these impacts can lead to increased consumer prices for insurance, food, water and energy – not paying attention in the short term can lead to higher costs for businesses and ordinary people in the longer term.

However, many businesses are now seizing the opportunities associated with putting the world on a pathway to low-carbon growth and development. Responding positively to the challenge could bring jobs and prosperity to communities.

Climate change in the distant past

The Earth's climate has always changed, with dramatic impacts for life on our planet.

Over millions of years the Earth has varied between extremes of almost completely frozen 'snowball' conditions and 'hothouse' states when forests have extended into the polar regions.

For the past 2.6 million years the Earth has always had some ice at the poles. However, the climate has moved repeatedly between colder and warmer periods in a cycle tied to the naturally recurring wobbles in the orbit of the Earth as it revolves around the Sun.

During the cold 'Ice Ages', large ice sheets spread over much of North America, Europe and Asia. They most recently reached their greatest extent about 22,000 years ago, when the sea level was some 130 metres lower than at present and the human population in Europe is estimated to have been about 130,000 – roughly the size of Cambridge, England today. In between the Ice Ages were warmer periods with much less ice and higher sea levels, during which different kinds of plants and animals flourished.

Severe and rapid climate fluctuations like those at the end of the last Ice Age probably contributed, along with hunting by humans, to the extinction of large mammals such as the woolly mammoth.

Causes of recent climate disruption

The scientific evidence shows that the dominant cause of the rapid warming of the Earth's climate over the last half century has been the activities of people, which have increased the amount of greenhouse gases in the atmosphere.

More greenhouse gases – carbon dioxide, methane, nitrous oxide and various industrial products – mean more of the energy we receive from the Sun is prevented from escaping back to space. This warms the planet.

Fossil fuels – coal, oil and natural gas – were formed from prehistoric plants and animals that took carbon from the atmosphere over the course of millions of years. As the fossil energy is burnt, that carbon is being released rapidly back into the atmosphere as carbon dioxide.

Clearance of forests, agriculture – including industrial farming methods that damage soils – and industrial processes such as cement production have also contributed significantly to an increase in greenhouse gas pollution.

Carbon dioxide levels worldwide are now more than 40 per cent higher than in 1750, at the start of the Industrial Revolution, when they were about 280 parts per million. By June 2016, even the remote Halley Research Station in Antarctica – far from sources of pollution – had recorded carbon dioxide levels surpassing 400 parts per million.

Halley Research Station

CO$_2$ 400 parts per million

Emissions, sinks, and atmospheric accumulation of carbon

Human activities emit about 36 billion tonnes of carbon dioxide into the atmosphere each year. The majority comes from burning fossil energy and industrial processes and the remainder is due to deforestation and other land-use changes.

Carbon dioxide is exchanged continually between the atmosphere, plants and animals through growth, death and decay, and also directly between the atmosphere and ocean. About half of the carbon dioxide pollution is soaked up by land and forests (land sink) or by the oceans (ocean sink). The rest, about 16 billion tonnes each year, accumulates in the atmosphere; there is no convenient hole in the sky for it to escape through. Without the natural sinks, this accumulation and the associated greenhouse warming would be greater and climate change would be even more severe.

The world warmed by about 4°C to 5°C over the millennia since the last Ice Age, as part of the natural cycle tied to changes in the Earth's orbit about the Sun. This drastically changed the conditions on Earth.

If the present accumulation of carbon dioxide in the atmosphere continues unchecked, greenhouse warming of similar magnitude might occur by as soon as the end of this century – this scale and speed of change would pose major challenges to human and natural systems.

Atmosphere
16 billion tonnes CO$_2$

9.5 billion tonnes CO$_2$
Ocean sink

Fossil fuels & industry
33 billion tonnes CO$_2$

10.9 billion tonnes CO$_2$
Land sink

Deforestation and land-use change
3.4 billion tonnes CO$_2$

Increasing energy demand

Total global energy use, including all domestic and industrial usage, has increased twenty-fold since 1850. This growth was accompanied by a shift from traditional energy sources such as wood, wind and water power towards fossil fuels, first coal and then oil and natural gas, as industrialization transformed the world.

Today fossil fuels make up almost 80 per cent of the world's energy use. Hydropower, wood, biofuels made from plants, and nuclear energy together account for just under 20 per cent. New renewable energy sources, such as solar and wind, represent about 2.5 per cent but are growing rapidly.

Providing clean, secure and affordable energy to everyone is one of the greatest challenges of the twenty-first century, as population increase and economic growth cause a rapid rise in demand for energy.

More than a billion people worldwide still live without access to electricity, mostly in Africa and Asia. Some three billion rely on wood or other solid fuel for cooking, or kerosene for lighting, resulting in indoor air pollution that causes millions of deaths each year. Outdoor pollution from burning coal and oil in power plants, industrial facilities and vehicles causes millions more deaths.

Clean renewable energy solutions offer ways to meet energy needs without the adverse effects of pollution on climate and health.

Clearing forests and damaging soils

Forests take carbon dioxide from the air and store it in trees, plants and soils. When trees are cut down and soil disturbed, much of that carbon is released back into the atmosphere.

For thousands of years, humans have cleared forest land to make way for farming. This deforestation was initially most widespread in the temperate regions of our planet. It not only released carbon, but also contributed to the extinction of some animals, such as tarpans (wild horses) and aurochs (wild cattle).

During the last century deforestation has accelerated in the tropical regions, including the equatorial rainforests. Some of these forests are located on areas of peat soils that contain vast quantities of carbon that accumulated slowly over many thousands of years.

When forests are cleared the soil may be damaged by erosion, and agricultural practices can deplete the soil's organic matter, releasing more carbon into the atmosphere.

The clearance and degradation of forests continues, mainly to make way for industrial farming. Among the industries with the biggest impact are palm oil, soya and beef. Deforestation is responsible for at least a quarter of carbon dioxide pollution over the past 150 years. Cutting down forests disrupts the water cycle, causing worsening drought in some regions. It also threatens many more animals and plants with extinction, including orang-utans, gorillas, tigers and forest rhinoceroses.

Past and present changes in carbon dioxide

Today's carbon dioxide levels now vastly exceed those reached at any point during at least the past 800,000 years, a period which extends deep into the Early Stone Age. This is revealed from ice cores drilled from the Antarctic ice sheets which present a clear record of the Earth's past climate and provide the most compelling evidence that changes over recent decades lie far outside the natural cycle.

As snow fell, over time it piled up, layer-upon-layer, year-after-year, trapping within it tiny bubbles of air. By drilling down through the ice to recover these frozen bubbles from depths of more than 3,000 metres, samples of the past atmosphere can be analysed.

The sediment at the bottom of the oceans can tell us about the even more distant past. This shows we are emitting carbon dioxide into the atmosphere about ten times faster than any natural release in the past 50 million years, at least.

Most of today's carbon dioxide pollution will remain in the atmosphere for decades or centuries, and some even for thousands of years, meaning the climate impacts will persist long into the future. Eventually the carbon will be captured and stored by natural processes, but at a rate that is much too slow to keep up with the current rates of pollution.

800,000 years of CO$_2$

CO$_2$ (parts per million by volume)

Hundreds of thousands of years ago

Acid oceans

As a further consequence of carbon dioxide pollution the oceans are rapidly becoming more acidic. Comparable rates of ocean acidification have not been seen for many millions of years, perhaps not since about 250 million years ago, when the biggest mass extinction of species took place.

About 30 per cent of the carbon dioxide we produce today is soaked up by the oceans. This reduces the amount accumulating in the atmosphere and limits the level of global warming, but it also increases the acidity of seawater.

Scientists are starting to understand how ocean acidification directly affects ecologically vital species, including oysters, clams, urchins, corals and various plankton, and may indirectly affect fish populations. A decline of these and other organisms could cause disruption to entire food webs. This would have devastating consequences for coastal communities dependent on fishing industries and bring knock-on effects for food supply.

Corals are severely affected by both warming seas – which cause bleaching – and ocean acidification. Combined with other factors such as storm damage, poor water quality and predation, this is leading to concerns for the future of natural wonders such as Australia's Great Barrier Reef.

Risk of major environmental changes

As the world warms up, there is an increased risk of potentially irreversible environmental changes or abrupt shocks occurring.

Even modest temperature rise may threaten the vast ice sheets covering Greenland and West Antarctica and lead to seas eventually rising by several metres, transforming global coastlines. Melting has been seen across more than half the Greenland ice sheet during some recent summers, and there is some evidence that the collapse of the ice sheet in West Antarctica may already be unstoppable.

Changed rainfall patterns can lead to forests drying out and increased fire risk. Fire and drought lead dense forests to change into more sparsely wooded savannas, or even grasslands. These store less carbon, which leads to increased warming and further rainfall changes. Fears have been raised that, in the extreme, this vicious circle could even trigger the rapid dieback of the Amazon rainforest.

Other examples of abrupt changes that could possibly occur in the future include mega-droughts, monsoon failures and the collapse of the ocean circulation associated with the Gulf Stream. Another concern is the potential for massive release of methane from the thawing of vast frozen stores in the Arctic, which would lead to further warming, especially since methane is many times more powerful as a greenhouse gas than carbon dioxide.

Benefits of limiting warming to 1.5°C

Across every continent and ocean, impacts on the natural world and human wellbeing from changes in climate have been seen in recent decades.

In 2016 the global average temperature was about 1°C warmer than in pre-industrial times. The warmer the world becomes, the graver the risks posed to the environment, human societies and the economy.

Climate change is a major concern for small island states such as the Maldives, Marshalls, Kiribati and Tuvalu. They worry about being engulfed by rising waters since many of the islands are only a couple of metres above sea level.

There is evidence that during an extended warm period 400,000 years ago a large fraction of Greenland was ice-free and sea levels rose slowly over centuries to be more than six metres higher than today. Local conditions influence the ice sheets, but at the time the global average temperature was perhaps only slightly warmer than today. This emphasizes the need to limit warming to avoid major long-term changes.

In the 2015 Paris Agreement the world's governments committed to hold the global temperature to well below 2°C above pre-industrial levels and to pursue efforts to limit the warming to 1.5°C by reducing and eventually eliminating greenhouse gas pollution.

Such action will not only reduce the overall risks posed by climate change but will also provide a range of other benefits. For instance, tackling climate change could be a huge opportunity to improve health worldwide through reducing pollution and encouraging healthy lifestyles.

Even modest warming
may threaten the
Greenland ice sheet

1.5 to
stay alive

The scale and urgency of pollution cuts

If the world is to meet the objectives of the Paris Agreement to limit climate change then carbon dioxide emissions will need to be eliminated in net terms by about the middle of this century. This means we need to cut polluting activities urgently.

The amount of carbon dioxide that can be released before dangerous levels of warming are reached can be seen as a carbon budget. Like a household budget, without careful planning it can be blown in wasteful ways.

The more carbon dioxide pollution we generate now, the faster we will have to reduce our emissions later to stay within the budget – and we may find that the speed of cuts then required is unachievable, even with new technologies.

The entire budget of carbon dioxide emissions to stay below 2°C of warming is about 3,000 billion tonnes, but it is estimated that our emissions since 1870 already amount to 2,000 billion tonnes. As a measure of the pressing urgency, at the present rate of fossil fuel use, deforestation and soil damage we are on course to exhaust the budget for 2°C within the next twenty to thirty years, with the budget for 1.5°C of warming being exhausted even sooner.

Moreover, even if we eliminate emissions, elevated temperatures will persist and some climate changes will continue to develop for many hundreds of years as, for instance, the deep oceans slowly warm, the ice sheets shrink and sea levels rise in response to the accumulated pollution.

Carbon budget for 2°C of warming

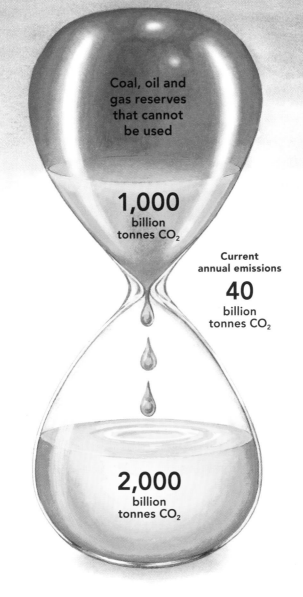

Coal, oil and gas reserves that cannot be used

1,000 billion tonnes CO_2

Current annual emissions **40** billion tonnes CO_2

2,000 billion tonnes CO_2

Energy solutions

To achieve the necessary cuts in greenhouse gas emissions we must take control of our energy future and make sensible adjustments to stop our reliance on fossil energy.

Avoiding waste and pollution by modernizing and shifting to renewable energy sources is vital. Abundant and secure clean energy supplies can be derived from the wind, sun, plants, tides, waves, modern waste-to-energy technologies and hot rocks underground. Nuclear power also produces low greenhouse gas emissions, but it is expensive.

Renewable energy will need to be developed in tandem with different kinds of electricity storage technologies, including new generations of batteries. Better batteries will also help shift our transportation to electric vehicles. To limit climate change, technologies must also be developed and adopted to capture carbon dioxide from power stations and store it underground.

Improvements in energy efficiency will be vital to cutting emissions quickly enough. New technologies including smart meters in homes and factories – joined with smart grids – will enable us to meet a bigger share of our needs from renewable energy sources and to do that in very efficient ways.

By later this century, we will need to have found ways of removing large amounts of carbon dioxide from the atmosphere without damaging side-effects, especially if we are to limit temperatures to a 1.5°C increase.

Forest solutions

Stopping deforestation, protecting natural forests, and restoring some of the forests that have already been lost – including on degraded lands that are unsuitable for farming – will be a major part of the solution to reducing greenhouse gas pollution and limiting climate change.

This is a complicated and fraught task, but when ways are found to make these forests valued it is possible to achieve progress. For example many countries are realizing that their tropical rainforests contribute to national development. They support sources of rainfall and clean water that sustain farming and the rivers that power hydroelectric dams. Forests also help to reduce flooding and protect soils.

These natural processes are essential for maintaining secure supplies of food and water. Therefore the protection of forests not only helps us to reduce carbon dioxide pollution, but also provides a wide range of other benefits.

Cocoa farmers in West Africa are being assisted to raise crop yields so they can increase their income without cutting down more forest. This includes switching to agroforestry, whereby crops and trees are grown together in ways that produce much more food from the same land. Food supplies are protected, and so are the exports that support the economy, while carbon remains locked up in the remaining rainforests.

Food and farming solutions

Farming can also be part of the solution, if food producers switch to methods which build up carbon levels in soils. Soils – especially those with a high proportion of organic matter – are important natural sinks of carbon.

Organic matter is made up of once-living material, such as dead plant leaves and roots, which are gradually broken down by tiny organisms in the soil, releasing nutrients that enable crops to grow without the need for manufactured fertilizers.

Increasing the amount of organic matter in the soil can best be achieved by using crop rotations and by mixed farming systems which combine both crops and livestock. These and other techniques, such as applying composted animal and plant wastes and adding charcoal, reduce farmers' reliance on artificial fertilizers, the production of which uses fossil energy and releases high levels of nitrous oxide, a powerful greenhouse gas.

Soils with a higher proportion of organic matter also hold more water, helping to protect crop plants from the effects of drought, in turn making farming more secure as the climate changes.

Emissions can also be reduced by cutting food waste and through diets that include less, but better quality, meat and dairy products.

Circular economy

Our planet and its ecosystems run through cycles and loops, for example the water cycle and carbon cycle. Soils break down plant remains and turn them into the nutrients needed to grow new plants. As is common sense, everything is recycled and reused: in Nature there is no waste.

In modern societies, however, our way of doing things is often more in straight lines than loops. We take resources, make products, use them and often dispose of them in waste that goes into the land, the atmosphere and the oceans without recapturing the resources used to make them.

This not only wastes the Earth's limited resources, such as aluminium and iron, but also uses up energy and leads to higher pollution than if we did things in more natural, circular ways. By harnessing new technologies, designing products differently and planning for a zero-waste future we could create a 'circular economy'.

The circular economy would copy Nature and do away with the idea of waste, instead seeing everything that we presently dispose of as a resource with which to make new products. Companies in some sectors are making huge strides in turning these ideas into reality, including the manufacture of fuels from industrial residues that arise in paper-making and which would otherwise be wasted.

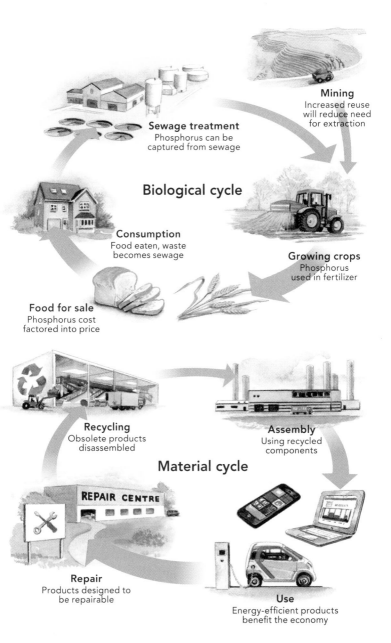

Sewage treatment
Phosphorus can be
captured from sewage

Mining
Increased reuse
will reduce need
for extraction

Biological cycle

Consumption
Food eaten, waste
becomes sewage

Growing crops
Phosphorus
used in fertilizer

Food for sale
Phosphorus cost
factored into price

Recycling
Obsolete products
disassembled

Assembly
Using recycled
components

Material cycle

Repair
Products designed to
be repairable

Use
Energy-efficient products
benefit the economy

REPAIR CENTRE

Meeting the challenge of climate change

'If, at last, the moment has arrived to take those long-awaited steps towards rescuing our planet and our fellow man from impending catastrophe, then let us pursue that vital goal in a spirit of enlightened and humane collaboration.'

HRH The Prince of Wales opening the UN Climate Change Conference in Paris, December 2015

Climate change poses a truly global challenge which no one country on its own can solve and which must be addressed together with sustainable development and efforts to eradicate poverty.

Businesses, leaders from major faith groups and millions of people around the world called for action. In 2015, virtually all of the world's countries pledged in the Paris Agreement to limit future warming through drastic cuts in fossil energy use and also by measures such as conserving and restoring forests and maintaining the health of our soils.

The pledged emissions reductions are not yet enough to hold temperature increases below 2°C, let alone 1.5°C, but there is a mechanism for review and strengthening of them.

The elimination of greenhouse gas emissions over the next few decades will limit future climate change, but some impacts on people, livelihoods and ecosystems will nevertheless still be felt around the world. In recognition of this, the Paris Agreement also set out the need to adapt to some climate change so that communities and countries will be more resilient and less vulnerable.

Nations Unies

Conférence sur les Changements Climatiques 2015

Paris France

PEOPLE'S CLIMATE MARCH

New opportunities and improved quality of life

Tackling climate change presents a huge opportunity for companies to develop new products, and in the process create new trading relationships and new jobs. It could also improve the quality of people's lives in many different ways.

Renewable energy technologies are growing fast, creating employment in manufacturing, installation and maintenance. Ultra-low-energy homes are being built that are comfortable and cheaper to run than standard designs, and incorporate innovative features such as solar roof tiles.

In transport, all major car manufacturers are developing electric vehicles. A new generation of liquid biofuels, for example made from algae or agricultural waste, could significantly reduce carbon pollution from transport, including from aircraft.

An increasing number of companies that make products with agricultural or wood content have adopted 'zero deforestation' policies. By using only sustainable sources the remaining natural forests can be better protected.

A new Industrial Revolution is underway which aims to meet people's needs without greenhouse gas pollution. Rapidly growing new business sectors are worth hundreds of billions of dollars per year. Governments can help accelerate this positive change by providing incentives to reduce emissions. For example, instead of subsidising fossil energy, they could require carbon polluters to pay for emissions, thereby encouraging investment in clean alternatives that create local jobs, stronger communities and a more stable climate.

CERTIFIED ULTRA-LOW-ENERGY HOME

One Earth

Over the centuries our demands on Nature have expanded massively, but there is still only one Earth.

Upon that Earth our impact grows as our population increases and as we consume more of its resources. If our society is to thrive and prosper long into the future without disastrously breaching Nature's boundaries, we must curb the growth in consumption and find ways of living in harmony with the world that sustains us.

We are already experiencing the impacts of climate change. Our children and grandchildren risk having to confront climatic dangers of our creation. It is vital we see the perils at hand and act accordingly before we test our world to destruction.

Future generations will have to live with the consequences of the choices we make over the coming years. Each of us must take responsibility – individually and collectively – to make those choices informed and positive, so we set out on a path to a secure and prosperous future founded on a healthy natural environment.

We have been mistreating our planet and abusing its natural systems. If the planet were a patient, we would have attended to her long ago. We have the power to put her on life support, and we must surely start the emergency procedures without further procrastination.

If we look after the Earth, only then will the Earth be able to look after us.

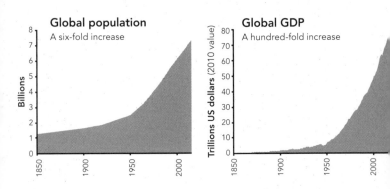

Global population
A six-fold increase

Billions

8
7
6
5
4
3
2
1
0

1850   1900   1950   2000

Global GDP
A hundred-fold increase

Trillions US dollars (2010 value)

80
70
60
50
40
30
20
10
0

1850   1900   1950   2000

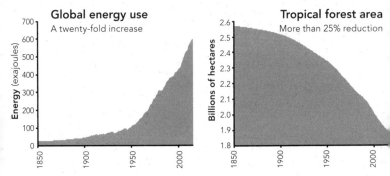

Global energy use
A twenty-fold increase

Energy (exajoules)

700
600
500
400
300
200
100
0

1850   1900   1950   2000

Tropical forest area
More than 25% reduction

Billions of hectares

2.6
2.5
2.4
2.3
2.2
2.1
2.0
1.9
1.8

1850   1900   1950   2000

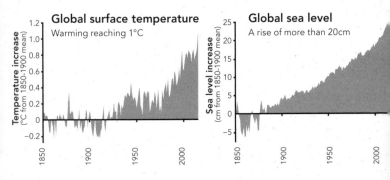

Global surface temperature
Warming reaching 1°C

Temperature increase (°C from 1850-1900 mean)

1.2
1.0
0.8
0.6
0.4
0.2
0
-0.2

1850   1900   1950   2000

Global sea level
A rise of more than 20cm

Sea level increase (cm from 1850-1900 mean)

25
20
15
10
5
0
-5

1850   1900   1950   2000